10 Simple Tricks of Mental Math

Reuven Rosenfeld

About the Author

Reuven Rosenfeld is a Chinese American entrepreneur, writer, startup employee and artificial intelligence software developer. In real life, he is an employee of a high growth startup company focused on hardware-based artificial intelligence. He has extensive research and development experience in software industry, including two universities, two startups and one industrial research institute. He has a Master degree in Electrical and Electronic Engineering.

Reuven learned his mental math skill since he was young. Among friends, he was always designated to be the person who splits the bill, check arithmetic calculation and generally …… calculate.

"10 Simple Tricks of Mental Math" is his second book. Reuven also wrote extensively about poker under the name "Jeremy Stryker".

Foreword

As a lover of Math and arithmetic, I have been into mental math since I was young. I read many great books of mental math such as "Short-Cut Math" and "How to Calculate Quickly: Full Course". In real-life, I have been blessed with capability of doing complex arithmetic on the fly. This has been a great help of my life. With mental math, I was able to navigate through a lot of life situations. For example, how much should you pay in tips? Say if an investment is 8%, when would you be able to double up your investment? For the unknowing, mental math is a kind of black magic.

Yet, I believe everyone can learn mental math. In this short book, I am going to share 10 simple rules on how mental math works in practice. Unlike many other books, I will not burden you with tons of exercises. Rather I will give you several tricks you can use right there.

For some tricks, I will teach you method that you can calculate quickly. In computer science, those are called the "algorithm". e.g. Trick 1 "Squares of number ended with 5" gives you an impressive party trick such that you can calculate the squares of two or even three digit number instantly.

In some cases, I tell you the principles of mental math. Why sometimes certain rules take a long time to execute. e.g. Trick 5 "Always working out the largest digit first in addition". This allows you to call of or understand seemingly "smart" mental mather. The truth, not many of them are that smart once you understand their trick.

In any case, sit back and I hope you enjoy.

Reuven Rosenfeld

July 2013

Trick 1 : Square of a number ended with 5

A square of a number is simply the number multiplied by itself. For example, the square of 5 is 5 x 5 = 25. Whereas, the square of 71 is 71 x 71 = 5041.

Most of us know from multiplication tables how to calculate the squares from 1 to 10.
So ......

Square of 1 is 1x1 =1
Square of 2 is 2x2 =4
Square of 3 is 3x3 =9
Square of 4 is 4x4 =16
Square of 5 is 5x5 =25
Square of 6 is 6x6 =36
Square of 7 is 7x7 =49
Square of 8 is 8x8 =64
Square of 9 is 9x9 =81
Square of 10 is 10x10 =100

But what if the number is bigger than 10? It starts to get more difficult. For example, you *might* know square of 11 is 121. But how about square of 29? (Ans: 841)

Now, square turns out to be a very important tool in mental math so there're some benefits of remembering all the square number (See Trick 3: "Remember the Squares").

Not all squares are difficult. In some cases though, square a number can be worked out in your head! Here is what we are getting at in this trick.

For any two digits number which end at 5. Do the following
Step 1, Find the tenth digit, call it a.
Step 2, Calculate $b = a \times (a + 1)$
Step 3, Put 25 behind b, then you got the answer!

Here is an example. Say the number is 85.
Step 1, Find the tenth digit. It's 8.
Step 2, Calculate $b = 8 \times (8 + 1) = 72$
Step 3, Put 25 behind b, i.e. 72. So you have 7225.

And you can use a calculator to verify it. 85 x 85 is indeed 7225.

Now, you can use the same trick for numbers such as 35, 45, 75. (Ans: 1225, 2025, 5625.)

Simple?

It is. You can even use it for number such as 115. In that case, the "tenth-digit" is 11. b = 11 x (11 + 1) = 132. So square of 115 is 13225.

If you are curious of why this trick always work, here is an algebraic explanation.

For a number n = (10a + 5), n x n = (10a+5) x (10a+5) = 100 a x a + 100 a + 25
or 100 a (a + 1) + 25. That explain why this tricks work all the time.

Trick 2: Square of a number ended with 1

After Trick 1, what about square of other numbers? Is it possible to calculate them quickly.

Let me tell you one more, which is the square of any two-digit number ended in 1. e.g. 71, 81, 11. etc.

So what you need to do is
1, identify the tenth digit.
2, Square the tenth digit and put it in the hundredth place.
3, Double up the tenth digit and put it in the tenth place.
4, Add one at the unit place.

e.g. 71
1, The tenth digit is 7.
2, Square the tenth digit, we got 49, and we put it in the hundredth place.
3, Double up the tenth digit, we got 14, and we have 490 + 14 = 504.
4, Add one at the unit place. So we have 5041.

This one is slightly tougher. In fact, it's very close to direct multiplication we learn in school but it's still slightly simpler.

81? 6561. 11? 121. 111? (!) 12321.

Trick 3: Remember the Squares

"Okay now." You say. "Those are simple tricks. If we don't have a number which ends with 1 or 5, does it mean we are stucked?"

In a way it is, you might need to resort to direct multiplications to calculate square. e.g. square of 83. You need to

```
  83
  83
-----
 249
664
------
6889
```

This is slow. So what is a better way?
So here is a dirty trick: You *remember* the squares. For starter, try to remember the squares from 1 to 25.

They are

Square of 1 is 1x1 =1
Square of 2 is 2x2 =4

Square of 3 is 3x3 =9
Square of 4 is 4x4 =16
Square of 5 is 5x5 =25
Square of 6 is 6x6 =36
Square of 7 is 7x7 =49
Square of 8 is 8x8 =64
Square of 9 is 9x9 =81
Square of 10 is 10x10 =100
Square of 11 is 11x11 =121
Square of 12 is 12x12 =144
Square of 13 is 13x13 =169
Square of 14 is 14x14 =196
Square of 15 is 15x15 =225
Square of 16 is 16x16 =256
Square of 17 is 17x17 =289
Square of 18 is 18x18 =324
Square of 19 is 19x19 =361
Square of 20 is 20x20 =400
Square of 21 is 21x21 =441
Square of 22 is 22x22 =484
Square of 23 is 23x23 =529
Square of 24 is 24x24 =576
Square of 25 is 25x25 =625

It will be better to memorize the square table from 1 to 100.

Memory? Aren't we go to *calculate* the product?

Here is one confusion of beginner of mental math. They refuse to believe that memory is part of the resource you use in fast mental math. More often than not, memory is *crucial* in fast mental math.

"Woa, it's difficult to remember so much."

You are sort of right. If you really want to use this memory trick in mental math, instant memory is very important. For example, when you see 24, you should be able to come up with 576 instantly.

Why is it only "sort of right"? The reason is there is a pattern in squares. Say for suppose, you remember the square of 22, which is 484. What is the square of 23? The answer is 484 plus 2 times 22 plus 1. That is 529.

In general, say you know the square of a, let's call it b. The square of (a+1) is b + 2 x a +1. This aids your memory.

Trick 4: Now I know the Squares, then what?

Good questions. So first of all, *you already know how to calculate the squares.* That is already a very impressive party trick.

But what else does it bring you?

Consider the calculation 29 x 33, in which the numbers are either both odd or both even. Knowing the squares will allow you to calculate the product quickly.
1. Find the mid point of the two numbers, call it a.
2. Find the difference between the number and the mid point, call it b.
3. Calculate square of a.
4. Calculate square of b,
5. Calculate the difference between square of a and square of b. That is the answer.

In our example, 29 x 33.
1. The mid point is a = (29 + 33) / 2 = 31
2. The difference is b = 31 - 29 = 2
3. Square of a = 961

4. Square of b = 4
5. The difference is 961 - 4 = 957. That is the answer.

Why this this work. In general, when you calculate multiplication, you can always make the product in the form of (a + b) (a - b) where a is the midpoint of the two numbers and b is the distance from the mid-point. But (a + b) (a - b) happens to be equal to a x a - b x b. Bingo. You got the answer.

If you already know the square table within 100, using the above tricks will allow you to calculate 50% of all multiplication between two 2-digit numbers. Very cost effective, isn't it?

Trick 5: Always calculate the Largest Digit in Addition

Let us go back to do something "simpler": addition. How do you usually do addition then? Say we want to add 128 + 429. What will be the answer?

If you follow the school teaching, you will write down 128, 429 like this

```
  128
+ 429
--------
  557
```

Now here is one bad habit in addition if you learn from high school. Usually they ask you to calculate from the unit digit. So 8 + 9 = 17. So the unit place is 7, we have a carry....... and so on so forth.

That is also slow. In fact, what you should do is the calculate from the largest digit. In this case, the hundredth digit. So you should have 1 + 4 = 5, ask "is there a carry?". No. say 5. 2 + 2 = 4, ask "is there a carry". Oh yes. So say 5 as well. And eventually you have 557. (Don't need to think of the carry in the unit place.)

Why do you want to do that? In fact there is no difference in terms of the number of calculation you do. But for most of the time, calculating the largest digit first is still a better habit, *why?*

The reason is that let's say if we just want to have an *approximate* answer for the sum. We can stop at any point we want. For example, consider the following sum

245989 + 356789.

You can calculate the whole thing. The answer is 602778. But if you go from the largest digit and go backward, you will be able to say. Oh the number is approximately 590000. 601000 …... 6027600 etc.

That's for most of the time, it's exactly what you want. Because of the time people only care the first few digit of your answer. If you can say the above sum is approximately 600 thousand, many people will feel happy enough with your mental math skill!

Trick 6: Calculate the Squares for a 3-Digit Number

Suppose you already know the squares from 1 to 100. How do you know the square of say 244?

Here is what you do.
1, Learn the hundredth digit, call it a. Call the rest (the tenth digit and onward) b.
2, Sum the number with b, call it c
3, Multiply c with *a* call it d.
4, Calculate the squares of b and add it to 100 times d. That's the final answer.

So for 244, for example.
1, the hundredth digit is 2, the rest is 44
2, Sum the number 244 by b = 44. i.e. c= 288.
3, Multiply c with a = 2. i.e. 576.
4, Calculate square of b, from memory you know squares of 44 is 1936. So the final answer is 100 x 576 + 1936. Or 59536.

Why does it work? The algebraic explanation is here. For a 3 digit number 100a + b. The square is

$10000 \, a \times a + 200 \times a \times b + b \times b.$
$= 100 \, a \, (a + 2b) + b \times b.$

That explains why the whole scheme works.

Trick 7: Don't do division, do multiplication

We all know the four basic operations of numbers: addition, subtraction, multiplication and division. Once you get into addition and subtraction seems to have similar difficulty. Multiplication, once you remember the multiplication tables, you seem to be in a good shape too. (Of course, knowing Trick 1 to 4 and 6 doesn't hurt neither.)

What stumble most people is division. How to do it well? Here is a general principle, don't go to do division, do multiplication and get close to the dividend.

For example, suppose you want to calculate 641 divided by 7. How do you do it?

You can approach by what school taught you.
```
    91
   ___
7 | 641
    63
```

```
   ___
    11
     7
   ___
    4.
```

You ended up get the answer 91 with a remainder of 4.

Or you can try to guess which multiple of 7 is close to 641. For me, once I saw 64, I think of 9, because 9 x 7 is 63, so the quotient got to have 90 there. But we still have 11 left. So it is another 1 we need.

Multiplication is fast for both human and computers. Remember that, you will calculate much faster.

Trick 8: Split the Bill Evenly (with tips)

One real life situation about division is how you should split the bill. For example, if you have 6 person and the bill total is 173.56. Without tips, this is a division exercise (See Trick 7). The answer is 28.92 with 4 cents left.

But what if you need to pay tips? In fact, this is a case where your division would greatly simplify. Why? Because you don't always need to give an exact percentage tip for people to feel grateful for you. The standard 15% rule can simply be "around 15%". This greatly simplifies the calculation.

In our case, this is how I think the calculation. $173.56's 15% is approximately like $170's 15%. or $28.5. So the total is $198.5. Take a convenient multiple of 6. Oh $198. Divide it by 6 is $33.

That is how you think in practice when you think of tip. Note, it is slightly smaller than 15% in practice but it's close enough no one would go to question you.

Rule of 72 and 69

Suppose you have an investment with growth is 4.5%. How long would it take for the investment to double up?

If you want to calculate this quickly, you would need to have a scientific calculator. But this rule of compounding has some real simple rule. If you rate is x%, you just need to calculate

72/x if you do simple compounding.

In our example, 72/4.5 = 16. So it takes 16 years to double up.

A related rule is rule of 69. That happens when you calculate interest continuously. In that case, it will take 69/4.5 = 15.33 years to double your investment up.

Out of all the tricks we discussed so far. This is perhaps the most practical. Always run this rule when you need to buy a house or invest!

Trick 10: Creative Mental Math

As you know, I haven't given a comprehensive treatment of the four operations. e.g. you can only calculate 50% of the situations for multiplications two 2-digit numbers. I also haven't discussed nuance of the four operations. What if we mix the operation together? What if we have 3 digit numbers? Those I plan to do it with another book. (say a "Comprehensive Mental Math from Calculo Rapio")

For a 99 cents book, I think I gave you several solid building blocks to improve your mental math. So before I go, let me say a thing or two about mental math.

Here is an interesting story. The great physicist Richard Feynmann once tested a mental math expert. Let me paraphase the question: 100 x 1789. The mental math to come up with the answer 178900. As you know, it is kind of silly. You just need to "shift 1789 for two digits and filled in two zeros", you will get the right answers.

So what's wrong? Most of mental math experts advocate that all calculations have to be done in the same way. In real life though, you need to be creative and apply several different rules to do calculations.

In the above example, If you calculate 100 x 1789 through complex rules. It will take you forever to come up 178900. But once you realize this is just 1789 x 1 x 100, you will come up with 178900 very quickly.

What's the lesson? Mental math, just like many disciplines in your life, require flexibility and creativity. It has never been just the memorization of rules. Rather how to use them at the right time is more important in practice.

In any case, I hope you enjoy this short book, until next time.

www.ingramcontent.com/pod-product-compliance
Lightning Source LLC
Chambersburg PA
CBHW070736180526
45167CB00004B/1774